CLIMATE CHANGE,

IS IT A *SCAM*?

DAVID R WISSON

TABLE OF CONTENT

Prologue

1. Introduction
- *Background and significance of studying temperature fluctuations*
- *Brief overview of methods and records used for temperature reconstruction*

2. Paleoclimatic Records
- *Overview of different paleoclimatic proxies (e.g., treerings, sediment cores, corals)*
- *Explanation of how paleoclimatic data are used to reconstruct past temperatures*
- *Case studies of notable paleoclimatic records (e.g., Fusel records, ice cores, lake sediments)*

3. Ice Core Records
- *Explanation of ice core drilling and analysis techniques*
- *Interpretation of ice core isotopic data (e.g., $\delta 18O$ and δD) for temperature reconstruction*
- *Notable ice core records and their insights into temperature fluctuations*

4. Historical Climate Data
- *Utilization of historical records (e.g., diaries, ship logs, agricultural records) to infer past temperatures*
- *Challenges and limitations of using historical data for temperature reconstruction*
- *Examples of key historical events influenced by temperature fluctuations*

5. Proxy Data Synthesis

*   *Integration of various proxy data sources for a comprehensive temperature reconstruction*
*   *Statistical and computational methods for combining diverse proxy records*
*   *Case studies showcasing the synergy of different proxy data sets*

6. Natural Climate Forcings

*   *Examination of natural factors influencing temperature fluctuations (e.g., solar variability, volcanic activity)*
*   *Discussion of how these factors are captured in the temperature records*
*   *Quantifying the relative contributions of different natural forcings*

7. Anthropogenic Influences

*   *Assessment of human activities impacting global temperatures (e.g., greenhouse gas emissions, land use changes)*
*   *Separating natural variability from anthropogenic signals in temperature records*
*   *Comparison of recent temperature trends with pre-industrial periods*

8. Climate Models and Simulations

*   *Introduction to climate models and their role in understanding temperature fluctuations*
*   *Calibration and validation of models using historical and proxy data*
*   *Model simulations of past temperature variations and comparisons with observations*

9. Impacts of Temperature Fluctuations

*   *Exploration of ecological, societal, and economic consequences of temperature changes*
*   *Examples of historical events illustrating the effects of temperature fluctuations*
*   *Consideration of future implications based on past temperature patterns*

10. Conclusion and Future Directions

- *Summary of key findings from the research*
- *Identification of remaining gaps and uncertainties in temperature reconstruction*
- *Suggestions for further research and exploration of potential applications*

PROLOGUE

In an era characterised by complex debates and conflicting narratives, few topics have evoked as much scrutiny and controversy as the issue of global warming. The urgency of this matter is underscored by the profound implications it holds for the fate of our planet and the well-being of future generations. The discourse surrounding global warming has given rise to divergent perspectives, ranging from vehement denial of its existence to urgent calls for immediate action. Amidst this cacophony of opinions, the objective pursuit of truth and scientific inquiry becomes ever more crucial.

This book embarks on a comprehensive journey into the heart of this debate, delving into the scientific underpinnings and historical context of global warming to determine whether the current crisis is an elaborate fabrication, or a stark reality grounded in empirical evidence. We traverse the annals of Earth's geological history, meticulously examining the intricate interplay of natural processes and anthropogenic influences to discern the veracity of the claims surrounding global warming.

The motivation behind this exploration stems from a collective awareness of the need to distinguish between unfounded scepticism and empirically substantiated truths. The rhetoric surrounding global warming has led to confusion and scepticism among the public, as well as fuelled controversies that have far-reaching implications for environmental policies, economic paradigms, and societal well-being. Thus, it is imperative to establish a nuanced understanding of the issue—one rooted in the robust scientific methodologies and historical context that have guided humanity's quest for knowledge.

Unravelling Earth's Climate History:

To address the central question of whether the current global warming crisis is a scam or substantiated by hard data, we embark on a multidisciplinary exploration that fuses insights from Earth sciences, atmospheric physics, historical climatology, and paleoclimatology. Our journey commences with a meticulous analysis of the Earth's climate history, as gleaned from an array of paleoclimatic records. These records, ranging from ice cores to sediment layers, provide windows into the past, allowing us to decipher the ebb and flow of temperatures and atmospheric conditions over millennia.

The examination of ice core records unveils a narrative that spans hundreds of thousands of years, revealing the intimate relationship between temperature fluctuations and the composition of Earth's atmosphere. The quantification of greenhouse gas concentrations—predominantly carbon dioxide (CO_2) enables us to discern the role of natural processes and human activities in shaping atmospheric dynamics. These insights, substantiated by data meticulously extracted from geological formations, guide us in understanding the inherent variability of Earth's climate and the drivers that propel it.

Beyond Natural Variability:

While the Earth's climate history elucidates the cyclical nature of temperature variations and atmospheric conditions, the recent surge in global temperatures has spurred debate over the extent to which human activities contribute to this phenomenon. The industrial revolution marked a turning point in human history, ushering in an era of technological advancement and fossil fuel consumption. Our analysis extends beyond the realm of natural variability to examine the exponential rise in CO_2 concentrations, unequivocally linked to human practices.

Through a meticulous evaluation of historical temperature records, we trace the trajectory of global warming in the modern era. Instrumental measurements collected from various sources—including weather stations, satellites, and oceanic observations—paint a coherent picture of the Earth's increasing surface temperatures. These temperature records, coupled with sophisticated climate models, enable us to differentiate between natural climate fluctuations and the marked influence of anthropogenic activities.

The Role of Scientific Consensus:

Central to our exploration is the role of scientific consensus as a guiding force in understanding the complex dynamics of global warming. The foundation of scientific inquiry rests upon rigorous examination, empirical validation, and robust peer review. This book places great emphasis on elucidating the overwhelming consensus among climate scientists regarding the reality of global warming and its connection to human activities. We draw from surveys and analyses that substantiate the consensus view, aligning with the positions of renowned scientific organizations such as the Intergovernmental Panel on Climate Change (IPCC), NASA, and NOAA.

Our endeavour delves into the methodologies employed by climate scientists to attribute temperature changes to specific drivers. We dissect the intricacies of climate modelling, a powerful tool that enables us to simulate past, present, and future climatic scenarios. These models, validated through historical data and paleoclimatic records, provide insights into the relative contributions of natural and anthropogenic factors to the observed warming trend.

Navigating Scepticism and Scams:

As we navigate the contours of this multifaceted discourse, we address the scepticism that often surrounds discussions of global warming. It is crucial to differentiate between genuine scientific inquiry and scepticism fuelled by misinformation or vested interests.
By providing a thorough examination of empirical evidence, historical context, and expert consensus, this book seeks to offer readers the tools to critically evaluate the claims, counterclaims, and controversies surrounding the global warming crisis.

In conclusion, the pages that follow endeavour to present an evidence-based exploration of the global warming crisis. Our journey through Earth's scientific and historical landscapes aims to ascertain whether the current state of alarm is founded on substantive data or if alternative explanations hold merit. With an unwavering commitment to scientific rigor and unbiased inquiry, we invite readers to join us in this voyage of discovery—a journey that seeks to illuminate the truth underlying the pressing global warming discourse.

CHAPTER 1

BACKGROUND AND SIGNIFICANCE OF STUDYING TEMPERATURE FLUCTUATIONS

Understanding the historical fluctuations in global temperatures is of paramount importance in deciphering Earth's climatic past, present, and future trajectories. By examining temperature variations over recorded history, we can gain insights into the natural rhythms of the climate system, identify potential drivers of change, and assess the impact of human activities on the environment. This knowledge is crucial for making informed decisions about environmental policies, resource management, and adaptation strategies.

Temperature fluctuations hold the key to unravelling complex interactions between various natural and anthropogenic factors that influence Earth's climate. Moreover, these fluctuations provide a baseline for evaluating the magnitude and pace of ongoing global warming due to human-induced greenhouse gas emissions. By contrasting past temperature changes with contemporary trends, we can distinguish between natural variability and anthropogenic influences, which is pivotal for projecting future climate scenarios.

Studying temperature fluctuations also offers insights into the resilience and adaptability of ecosystems and human societies. Historical temperature shifts have shaped agricultural practices, influenced the distribution of species, and even contributed to the rise and fall of civilizations. Lessons from the past can inform us about potential vulnerabilities and opportunities in the face of changing climates.

Brief Overview of Methods and Records Used for Temperature Reconstruction:

To reconstruct past temperatures, scientists employ a suite of proxy records, each offering a unique window into the climate of bygone eras. Proxy data are indirect indicators of climate conditions, encompassing a variety of sources such as ice cores, tree rings, sediment cores, corals, and historical documents.

1. Ice Core Records: Ice cores extracted from polar ice sheets and glaciers provide valuable insights into temperature variations and atmospheric composition. Isotopic ratios of oxygen and hydrogen within ice layers ($\delta 18O$ and δD) serve as proxies for temperature, revealing shifts in climate over thousands of years (Jouzel et al., 2007).

2. Tree Ring Analysis: Trees respond to changing climate conditions by altering their growth patterns. Dendrochronology, the study of tree rings, allows scientists to infer past temperature variations and even identify specific weather events (Esper et al., 2012).

3. Sediment Cores: Sediments deposited in lakes and oceans accumulate over time, preserving information about past climates. By analysing sediment layers, researchers can reconstruct temperature and precipitation patterns (Anderson et al., 2019).

4. Corals: Coral skeletons record variations in ocean temperature and chemistry. Geochemical analyses of coral growth layers provide insights into historical temperature changes in marine environments (Lough, 2007).

5. Historical Documents: Diaries, ship logs, and other historical documents offer anecdotal evidence of past climate conditions. These records provide localized snapshots of temperature fluctuations and their societal impacts (Pfister et al., 1995).

These proxy records are complemented by advances in statistical methods and computer simulations, which help integrate data from diverse sources and create comprehensive temperature reconstructions.

❖

CHAPTER 2

PALEOCLIMATIC RECORDS AND THEIR USE IN RECONSTRUCTING PAST TEMPERATURES

Overview of Different Paleoclimatic Proxies:

Paleoclimatic proxies are natural indicators that provide valuable information about past climate conditions. These proxies offer insights into temperature, precipitation, and other environmental variables. Different proxies are sensitive to various aspects of the climate system and are found in diverse natural archives. Some notable paleoclimatic proxies include:

Tree Rings:

Tree rings are among the most well-known and widely used paleoclimatic proxies. Trees form annual rings as they grow, and the width of these rings can vary based on environmental conditions. In regions with distinct growing seasons, such as temperate and boreal forests, tree rings can serve as a record of past temperature and precipitation variations. Dendrochronology, the study of tree rings, allows researchers to reconstruct climate patterns over centuries to millennia (Cook et al., 2020).

Sediment Cores:

Sediment cores extracted from lake beds, ocean floors, and other aquatic environments hold a wealth of information about past climates. These cores contain layers of sediment that accumulate over time, trapping pollen, microorganisms, and other materials. Analysis of these layers provides insights into changes in vegetation, land use, and climate. The composition of sediments can be used to infer past temperature and precipitation patterns, as well as shifts in hydrology and ecosystem dynamics (Bradley, 2015).

Corals:

Corals are unique paleoclimatic proxies found in tropical and subtropical marine environments. Coral skeletons grow in layers, with variations in their chemical composition reflecting changes in ocean temperature and other environmental factors. By analysing the isotopic and elemental composition of coral skeletons, researchers can reconstruct past sea surface temperatures and infer the frequency of events such as El Niño-Southern Oscillation (ENSO) events (Lough, 2007).

Explanation of How Paleoclimatic Data Are Used to Reconstruct Past Temperatures:

Paleoclimatic data are used to reconstruct past temperatures through a process known as proxy-temperature calibration. This involves establishing a relationship between the proxy data and instrumental temperature records from modern times. Statistical techniques, such as regression analysis, are employed to quantify this relationship and create a calibration model. Once calibrated, the proxy data can be applied to time periods before instrumental records were available.

For example, in dendrochronology, researchers identify a correlation between tree ring widths and local temperature observations. By comparing the widths of modern tree rings with temperature data, they develop a mathematical equation that links ring width to temperature. This equation can then be applied to older tree rings to estimate temperatures during previous centuries.

Similarly, sediment cores can be analysed for specific minerals, isotopes, or microfossils that respond to temperature changes. By understanding the relationship between these proxy indicators and temperature, scientists can create models that allow them to reconstruct past temperature variations based on sediment core data.

Case Studies of Notable Paleoclimatic Records:

Ice Core Records:

Ice cores extracted from polar ice sheets, such as those from Greenland and Antarctica, provide unparalleled insights into past climates. The isotopic composition of ice cores, along with trapped air bubbles and impurities, offer a detailed archive of temperature changes, greenhouse gas concentrations, and volcanic activity over thousands of years. By analysing the layers within ice cores, researchers can reconstruct temperature variations and climate events with remarkable precision (EPICA Community Members, 2004).

Lake Sediments:

Sediment cores from lakes around the world provide valuable information about regional climate variations. For instance, cores from alpine lakes may contain evidence of glacial advances and retreats, while cores from lowland lakes might reflect changes in precipitation and temperature. By analysing sediment layers, researchers can reconstruct past temperature and hydrological changes, shedding light on the interactions between climate, ecosystems, and human activities (von Gunten et al., 2013).

Historical Documents:

Historical documents such as diaries, ship logs, and agricultural records offer a unique perspective on past climate conditions.
These records provide anecdotal evidence of temperature extremes, unusual weather events, and their impacts on societies. For example, ship logs from the Age of Exploration document sea ice extent and can be used to reconstruct past maritime climate conditions (Pfister et al., 1995). Such documents provide localized snapshots of temperature fluctuations and contribute to the broader understanding of historical climate variability.

In conclusion, paleoclimatic proxies play a crucial role in reconstructing past temperature variations. Tree rings, sediment cores, corals, and historical documents offer insights into climate changes over various timescales. Through calibration and statistical modelling, these proxy data can be translated into estimates of past temperatures. Case studies of ice core records, lake sediments, and historical documents highlight the diverse sources of paleoclimatic data and their significance in unravelling Earth's climatic history.

CHAPTER 3

ICE CORE RECORDS

Explanation of Ice Core Drilling and Analysis Techniques:

Ice core records offer a direct and remarkably detailed window into Earth's climatic history, primarily through the analysis of polar ice sheets in regions like Greenland and Antarctica. Ice core drilling is an intricate process that involves extracting cylindrical samples of ice from deep within glaciers and ice sheets.

1. **Drilling Process:** Ice cores are retrieved using specialized drilling equipment designed to maintain the integrity of the ice while avoiding contamination. The core is extracted in sections, each representing a distinct time period. These sections are carefully stored and transported to laboratories for analysis.

2. **Ice Layer Dating:** The layers within an ice core act as a chronological archive, much like tree rings. Researchers use visual cues, such as variations in ice colour and particle content, to distinguish annual layers. By counting these layers and analysing ice chemistry, scientists can establish a precise dating of the core.

Interpretation of Ice Core Isotopic Data for Temperature Reconstruction:

Isotopic analysis of ice cores, particularly the ratios of oxygen isotopes ($\delta18O$) and deuterium (δD), provides invaluable insights into past temperature variations and atmospheric conditions.

1. **Oxygen Isotope Ratios:** The isotopic composition of oxygen in ice reflects the temperature at the time of ice formation. Warmer conditions lead to lighter isotopes being preferentially evaporated and transported to polar regions. Consequently, higher $\delta 18O$ values in ice cores indicate warmer periods, while lower values suggest colder periods.

2. **Deuterium (Hydrogen Isotope):** Like oxygen isotopes, deuterium content (δD) in ice cores is influenced by temperature. The correlation between δD and temperature is established based on laboratory experiments and observational data.

3. **Paleoenvironmental Insights:** Ice cores provide a wealth of information beyond temperature. By analysing trapped air bubbles and impurities (like dust and volcanic ash) in the ice, scientists can reconstruct past atmospheric compositions, greenhouse gas concentrations, and even volcanic events.

Notable Ice Core Records and Their Insights into Temperature Fluctuations:

1. **Greenland Ice Cores:** Ice cores from Greenland, such as those from the Greenland Ice Sheet Project (GISP2) and the North Greenland Eemian Ice Drilling (NEEM) project, have revealed significant temperature fluctuations over the past millennia. These records show distinct warming during the Holocene Climatic Optimum (around 6,000 to 9,000 years ago) and subsequent cooling periods, including the Little Ice Age.

2. **Antarctic Ice Cores:** Ice cores from Antarctica, like the EPICA (European Project for Ice Coring in Antarctica) Dome C core, have provided a high-resolution climate record covering about 800,000 years. These records highlight the glacial-interglacial cycles and the close link between temperature and atmospheric carbon dioxide (CO_2) concentrations.

3. **Volcanic Events and Climate Impact:** Ice cores also capture evidence of volcanic eruptions, which inject aerosols into the atmosphere. These aerosols can temporarily lower global temperatures by reflecting sunlight. Notable events like the eruption of Mount Tambora in 1815 are recorded in ice cores, demonstrating the climate impact of such events.

In conclusion, ice core records offer an extraordinary and intricate archive of past temperature variations and climatic conditions. Through a combination of careful drilling, isotopic analysis, and interpretation, scientists can reconstruct temperature changes with remarkable precision. These records not only provide insights into Earth's climatic past but also help validate climate models and inform our understanding of current and future climate change.

CHAPTER 4

HISTORICAL CLIMATE DATA

Utilization of Historical Records to Infer Past Temperatures:

Historical records, ranging from diaries and ship logs to agricultural documents and artwork, offer a unique perspective on past climate conditions. While these records may not provide direct temperature measurements, they contain valuable indirect indicators that can help infer past temperatures and climatic patterns.

1. **Diaries and Observations:** Personal diaries often include descriptions of weather conditions, such as temperature extremes, snowfall, and unusual weather events. For example, the diaries of Thomas Jefferson and George Washington in the United States contain detailed weather observations that can be analysed to reconstruct historical temperature variations.

2. **Ship Logs:** Ship logs from maritime explorations and trade voyages contain notes on weather conditions, wind patterns, and sea ice extent. These records provide insights into temperature and climate conditions at sea and along coastlines (Wood, 2010).

3. **Agricultural Records:** Farmers have historically kept records of planting and harvesting dates, which are influenced by temperature and growing season length. Analysis of these records can reveal shifts in local climate patterns and temperature fluctuations (Pfister et al., 1995).

Challenges and Limitations of Using Historical Data for Temperature Reconstruction:

While historical records offer valuable insights, they come with challenges that must be carefully considered when reconstructing past temperatures.

1. **Subjectivity and Bias:** Historical records are often subjective and influenced by personal perceptions and biases. Weather descriptions may vary in detail and accuracy among different observers.

2. **Spatial and Temporal Coverage:** Historical records are unevenly distributed in time and space. They may be concentrated in specific regions or time periods, limiting the ability to create a comprehensive global temperature reconstruction.

3. **Data Homogenization:** Historical records may be influenced by changes in observation methods, instruments, and station locations. Proper homogenization techniques are required to account for these biases and ensure data consistency over time (Jones et al., 2012).

Examples of Key Historical Events Influenced by Temperature Fluctuations:

1. **The Little Ice Age (14th-19th centuries):** Historical records document a period of cooler temperatures known as the Little Ice Age. This era witnessed significant societal impacts, including crop failures, famines, and population displacements. Glaciers advanced in Europe, and the freezing of rivers, such as the Thames in London, became more frequent (Fagan, 2000).

2. **The Medieval Warm Period (10th-14th centuries):** Historical documents suggest a warmer period during the Medieval Warm Period. This period is associated with agricultural expansion, increased settlement in Greenland, and cultural shifts in societies around the world.

3. **1816 - The Year Without a Summer:** Historical records document the unusual weather and temperature anomalies of 1816, which was dubbed the "Year Without a Summer." This phenomenon was attributed to the aftereffects of the massive eruption of Mount Tambora in Indonesia in 1815, which injected volcanic aerosols into the atmosphere and led to widespread crop failures and food shortages (Stommel, 1983).

In conclusion, historical climate data, derived from a wide range of sources, play a crucial role in reconstructing past temperature fluctuations and understanding their societal implications. While these records provide valuable insights, challenges related to subjectivity, spatial coverage, and data quality must be carefully addressed. By combining historical data with other paleoclimatic proxies, such as ice cores and tree rings, researchers can create comprehensive temperature reconstructions that enhance our understanding of Earth's complex climate history.

CHAPTER 5

PROXY DATA SYNTHESIS

Integration of Various Proxy Data Sources for Comprehensive Temperature Reconstruction:

Proxy data synthesis involves combining information from diverse sources, such as ice cores, tree rings, sediment cores, and historical records, to construct a holistic picture of past temperature variations. This integration is crucial for creating reliable and accurate temperature reconstructions that cover different spatial and temporal scales.

1. **Complementary Insights:** Each proxy data source has its strengths and limitations. By combining multiple proxies, researchers can leverage the strengths of each source to compensate for individual weaknesses. For instance, tree rings may provide high-resolution data for recent centuries, while ice cores offer longer-term records.

2. **Spatial Coverage:** Different proxies are often available from various regions around the world. Integrating data from multiple locations enables the creation of regional and global temperature reconstructions, enhancing our understanding of large-scale climatic patterns.

Statistical and Computational Methods for Combining Diverse Proxy Records:

The integration of proxy data sources requires advanced statistical and computational techniques to account for variations, uncertainties, and biases within the data.

1. **Climate Field Reconstruction (CFR):** CFR methods aim to create a spatially complete representation of past climate variability by combining proxy data with spatial patterns of modern climate observations. Techniques like RegEM (Regularized Expectation Maximization) and EOF (Empirical Orthogonal Function) analysis are used to extract climate signals from noisy proxy records (Mann et al., 2008).

2. **Data Assimilation:** Data assimilation techniques, commonly used in climate models, merge proxy data with physical models to create consistent and physically plausible reconstructions. These techniques help fill gaps in proxy records and provide a more continuous view of past temperature variations (Tingley and Huybers, 2010).

3. **Uncertainty Quantification:** Combining diverse proxy records introduces uncertainties arising from diverse sources of variability and errors. Bayesian methods are often employed to quantify and propagate uncertainties, providing a range of temperature reconstructions (Hakim et al., 2016).

4. **Weighted Averaging:** Proxy records are often combined using weighted averaging methods, where each proxy's contribution to the reconstruction is determined based on its reliability and signal strength. Weighting can account for proxy-specific uncertainties and provide more accurate estimates of past temperatures (McShane and Wyner, 2011).

Advantages and Challenges:

Advantages:

1. **Enhanced Reliability:** Integrating multiple proxy data sources reduces the risk of relying on a single source's limitations or biases, leading to more robust temperature reconstructions.

2. **Improved Temporal Coverage:** Some proxies provide high-resolution data for recent centuries, while others offer longer-term records. By combining these sources, reconstructions can cover a wider range of time scales.

3. **Spatial Insights:** Proxy data from different geographic regions provide insights into regional climate patterns and their interactions, enabling the identification of large-scale climatic drivers.

Challenges:

1. **Proxy Inconsistencies:** Proxy records may not always agree with each other due to local variations or proxy-specific biases. Reconciling these inconsistencies requires careful consideration and statistical techniques.

2. **Uncertainty Handling:** Integrating multiple sources of uncertainty, including measurement errors and proxy limitations, is a complex process that demands sophisticated statistical methods.

3. **Temporal Misalignment:** Proxy records may have different temporal resolutions and dating uncertainties, requiring careful alignment and synchronization.

Conclusion:

Proxy data synthesis is an intricate endeavour that involves harmonizing diverse sources of paleoclimatic information to reconstruct past temperature variations. By integrating data from ice cores, tree rings, sediment cores, and historical records using statistical and computational techniques, researchers can create comprehensive and reliable temperature reconstructions. Despite the challenges, proxy data synthesis provides a powerful tool for unravelling Earth's climatic history and advancing our understanding of temperature fluctuations over time.

CHAPTER 6

NATURAL CLIMATE FORCINGS

Natural Factors Influencing Temperature Fluctuations:

Temperature fluctuations on Earth are influenced by a complex interplay of natural factors, some originating within our planet's atmosphere and others arising from external sources. Understanding these natural climate forcings is essential for deciphering the drivers of past temperature changes.

Solar Variability:

The Sun is the primary energy source for Earth's climate system, and variations in solar radiation can have significant implications for temperature fluctuations. Solar activity is characterized by cycles of sunspots and solar irradiance, which affect the amount of energy reaching Earth's surface. While these cycles have small impacts on total solar irradiance, they can lead to subtle but noticeable changes in climate.

1. **Sunspot Cycles:** The Sun undergoes an 11-year cycle of varying sunspot activity. During periods of high sunspot activity, the Sun emits more energy, leading to a slight warming of Earth's surface (Lean, 2000).

2. **Solar Irradiance:** Changes in solar irradiance, particularly in the ultraviolet part of the spectrum, can influence the stratosphere and upper troposphere, potentially affecting atmospheric circulation patterns and temperature profiles (Haigh, 2003).

Volcanic Activity:

Volcanic eruptions release vast amounts of ash, sulphur dioxide, and other aerosols into the atmosphere. These particles can reflect sunlight back into space, leading to a temporary cooling effect on the planet. The impact of volcanic activity on temperature fluctuations is most pronounced in the years following major eruptions.

1. **Aerosol Injection:** Volcanic aerosols can block a portion of incoming solar radiation, causing a reduction in surface temperatures for a few years after the eruption. The year without a summer in 1816, following the eruption of Mount Tambora in 1815, is a notable example of the cooling effects of volcanic activity (Robock, 2000).

2. **Sulphate Aerosols:** Sulphur dioxide released during volcanic eruptions reacts with water vapor to form sulphate aerosols, which can remain in the atmosphere for an extended period. These aerosols scatter sunlight, leading to a dimming effect and subsequent cooling (McCormick et al., 1995).

Orbital Forcing:

Earth's orbital parameters, known as Milankovitch cycles, influence the distribution of solar energy received by different latitudes and seasons. These cycles, driven by variations in Earth's axial tilt, eccentricity, and precession, have played a crucial role in shaping the timing and extent of glacial and interglacial periods.

1. **Axial Tilt (Obliquity):** Changes in Earth's axial tilt affect the intensity and seasonality of solar radiation received at different latitudes. Greater tilt leads to more extreme seasons and can influence the initiation of ice ages (Loutre and Berger, 2003).

2. The fascinating interplay between the Gas Giants, Earth's orbit, and their cumulative effects on Earth's climate over millions of years.

Eccentricity, Gas Giants, and Earth's Climate:

Introduction to Eccentricity and Gas Giants:

Eccentricity refers to the extent of deviation of an orbit from a perfect circle. Earth's orbit around the Sun is not a perfect circle but an ellipse, with its eccentricity changing over geological timescales. One of the intriguing factors affecting Earth's orbit is the presence of massive Gas Giants, specifically Jupiter and Saturn, and their influence on Earth's orbital dynamics.

Gas Giants and Earth's Orbit:

The Gas Giants, Jupiter and Saturn, are colossal celestial bodies with immense gravitational influence. These planets significantly affect the motion of other objects in our solar system, including Earth. The interactions between the Gas Giants and Earth's orbit can lead to subtle but significant changes in the eccentricity of Earth's orbit, which can have profound implications for Earth's climate over geological epochs.

Sun's Center of Mass:

The gravitational forces exerted by the Gas Giants on the Sun result in a shift in the Sun's centre of mass. This shifting centre of mass causes the Sun to move in response to the gravitational pulls of the Gas Giants, creating a wobbling motion known as the "solar barycentre wobble."

Effect on Earth's Orbit:

The gravitational tugs of the Gas Giants on the Sun's centre of mass cause a corresponding wobbling effect in the Sun's orbit around the solar system's centre of mass. This wobbling of the Sun's orbit influences the orbits of the planets, including Earth, potentially leading to variations in Earth's eccentricity.

Impact on Eccentricity:

The varying eccentricity of Earth's orbit, driven by the interactions with the Gas Giants, can influence the distribution of solar radiation received by Earth. Changes in eccentricity alter the intensity of sunlight at different latitudes and times of the year, potentially affecting climate patterns and, in turn, contributing to glacial-interglacial transitions.

Cumulative Effects on Earth's Climate:

The cumulative effects of the Gas Giants' gravitational influences on Earth's orbit have the potential to influence Earth's climate over millions of years. Here, we explore how these effects accumulate and contribute to Earth's long-term climate variability.

Orbital Resonance and Long-Term Cycles:

The gravitational perturbations induced by the Gas Giants can lead to orbital resonances, where the period of one celestial body's orbit becomes a simple fraction of another's. These resonances can induce gradual shifts in Earth's orbital parameters, including eccentricity, over vast timescales.

Milankovitch Cycles and Climate Forcing:

The cumulative changes in Earth's eccentricity, along with its axial tilt and precession, are collectively known as Milankovitch cycles. These cycles play a pivotal role in driving variations in Earth's insolation (incoming solar radiation), which can trigger climate changes and influence the timing of glacial-interglacial transitions (Huybers, 2006).

Mechanism of Climate Impact:

The influence of the Gas Giants on Earth's eccentricity modifies the distribution of solar energy received at different latitudes and times of the year. These changes can affect the strength of seasons and potentially contribute to shifts in global climate patterns, driving long-term climate variability.

Paleoclimate Evidence:

Paleoclimate records, including ice cores and sediment cores, provide invaluable insights into Earth's past climate variability and its connection to changes in orbital parameters. The analysis of these records has revealed correlations between eccentricity-driven variations in insolation and significant climate shifts (Hays et al., 1976).

In summary, the interaction between the Gas Giants, Earth's orbit, and their cumulative effects on climate over geological timescales underscores the intricate dynamics of our solar system. The gravitational influences of the Gas Giants, particularly Jupiter and Saturn, play a crucial role in shaping Earth's orbital characteristics, including eccentricity, which in turn influence the distribution of solar radiation received by our planet. Understanding these complex interactions offers a window into the long-term drivers of Earth's climate variability.

Discussion of How These Factors Are Captured in the Temperature Records:

Understanding how natural climate forcings are captured in temperature records requires examining proxy data from various sources.

Ice Core Records:

Ice cores extracted from polar ice sheets provide a detailed historical record of past climate conditions, including the influence of solar variability and volcanic activity. These records reveal layers of ash and sulphate corresponding to volcanic eruptions. Sulphate concentrations in ice cores can be linked to specific eruptions, and the cooling effects of these eruptions are evident in isotopic and chemical data (Sigl et al., 2015).

Tree Ring Data:

Tree ring records can reflect changes in solar activity and the cooling effects of volcanic eruptions. Reduced tree growth during volcanic winters, marked by reduced solar radiation, can be observed in tree ring widths. Solar activity can also leave imprints on isotopic ratios within tree rings, offering insights into historical solar variability (Gray et al., 2010).

Climate Models and Simulations:

Climate models, which simulate Earth's climate system, are powerful tools for understanding the influence of natural climate forcings on temperature fluctuations. By running model simulations with and without specific forcings, researchers can isolate and quantify their effects on temperature variability (Hegerl et al., 2007).

Quantifying the Relative Contributions of Different Natural Forcings:

Quantifying the relative contributions of various natural climate forcings to temperature fluctuations is a challenging but essential task in climate science.

Attribution Studies:

Attribution studies use climate models and observations to assess the extent to which observed temperature changes can be attributed to specific forcings. By comparing model simulations with observed temperature records, researchers can estimate the contributions of solar variability, volcanic activity, and other factors (Stott et al., 2016).

Volcanic Forcings:

Volcanic aerosols have a well-defined and distinct signature in temperature records. By analysing the timing and magnitude of volcanic eruptions, scientists can estimate the cooling effects of volcanic aerosols on global temperatures. This information can be used to quantify the impact of volcanic forcings on temperature fluctuations (Robock, 2000).

Advantages and Challenges:

Advantages:

1. **Insight into Natural Climate Dynamics:** Studying natural climate forcings provides valuable insights into the inherent variability of Earth's climate system, helping distinguish between natural and human-induced temperature changes.

2. **Long-Term Perspective:** Natural climate forcings offer a long-term perspective on climate variability, spanning geological timescales and providing context for recent changes.

Challenges:

1. **Complex Interactions:** Natural climate forcings often interact with each other and with human-induced factors, making it challenging to isolate their individual contributions to temperature fluctuations.

2. **Data Limitations:** Historical records and proxy data may have limitations in terms of spatial coverage, temporal resolution, and uncertainties, which can impact the accuracy of quantifying natural forcings.

Never-the-less, the examination of natural climate forcings, including solar variability, volcanic activity, and orbital forcing, is crucial for understanding the drivers of past temperature fluctuations. Through the analysis of proxy records, climate models, and attribution studies, scientists can decipher the relative contributions of these forcings to temperature variability. This knowledge not only enhances our understanding of Earth's climate history but also helps contextualize current and future temperature changes in the context of natural climate dynamics.

CHAPTER 7

ANTHROPOGENIC INFLUENCES

Human Activities Impacting Global Temperatures

The influence of human activities on global temperatures is a topic of paramount importance in the study of climate change. Key anthropogenic factors include the emission of greenhouse gases (GHGs) and land use changes.

Greenhouse Gas Emissions:

Human activities, particularly the burning of fossil fuels for energy and industrial processes, have significantly increased the concentrations of carbon dioxide (CO_2), methane (CH_4), and other GHGs in the atmosphere. These gases trap heat, leading to a "greenhouse effect" that warms the planet.

1. **Fossil Fuel Combustion:** The burning of coal, oil, and natural gas for energy releases substantial amounts of CO_2 into the atmosphere (IPCC, 2014).

2. **Deforestation and Land Use:** Changes in land use, such as deforestation and urbanization, alter the Earth's surface characteristics, affecting heat absorption and reflection (Bonan, 2008).

3. **Industrial Processes:** Other activities, such as cement production and deforestation, can release GHGs and aerosols into the atmosphere (Myhre et al., 2013).

Land Use Changes:

Alterations in land use, driven by agricultural expansion, urbanization, and deforestation, can significantly impact local and regional climates. These changes can influence temperature patterns through modifications in surface albedo, heat absorption, and the exchange of moisture between the land and the atmosphere.

Separating Natural Variability from Anthropogenic Signals:

Distinguishing between natural climate variability and anthropogenic signals is a complex challenge. Robust methods are required to isolate the influence of human activities on temperature records.

Climate Models and Attribution Studies:

Climate models play a crucial role in separating natural variability from anthropogenic signals. By simulating Earth's climate with and without human influences, researchers can compare model outputs with observed temperature records to quantify the impact of anthropogenic factors (Stott et al., 2000).

Detection and Attribution:

Detection and attribution studies assess the likelihood that observed temperature changes can be attributed to specific drivers, such as human activities or natural processes. These studies utilize statistical analyses and simulations to attribute temperature trends to different factors (Hegerl et al., 2007).

Comparison of Recent Temperature Trends with Pre-Industrial Periods:

Assessing recent temperature trends in the context of pre-industrial periods provides insights into the magnitude and pace of anthropogenic climate change.

Historical Temperature Records:

Historical temperature records, such as instrumental measurements and historical documents, offer insights into temperature variations before the industrial era. These records enable the comparison of recent warming trends with past temperature fluctuations (Jones et al., 2012).

Proxy Data and Paleoclimate Records:

Proxy data, including tree rings, ice cores, and sediment records, extend temperature reconstructions back in time. These records provide a longer-term perspective on temperature changes and help contextualize recent warming within a broader historical context (Mann et al., 2008).

Anthropogenic Warming Signal:

Comparing recent temperature trends with pre-industrial periods reveals a clear anthropogenic warming signal. The rate and extent of temperature increase observed in the industrial era exceed natural variability, highlighting the role of human activities in driving recent climate change (IPCC, 2021).

Therefore, anthropogenic influences on global temperatures, characterized by greenhouse gas emissions and land use changes, have become dominant drivers of climate change in recent decades. Distinguishing human-caused changes from natural variability requires sophisticated modelling and attribution studies. Comparing current temperature trends with pre-industrial periods provides compelling evidence of the substantial impact of human activities on Earth's climate system.

CHAPTER 8

CLIMATE MODELS AND SIMULATIONS

Introduction to Climate Models and Their Role:

Climate models are sophisticated computer-based tools that simulate Earth's climate system, capturing the interactions between various components such as the atmosphere, oceans, land surface, ice, and even human activities. These models are essential for gaining insights into the complex dynamics of temperature fluctuations and their underlying mechanisms.

Components of Climate Models:

Climate models are composed of mathematical equations representing fundamental physical, chemical, and biological processes. These equations are solved numerically to simulate how different components of the climate system interact and evolve over time.

Role of Climate Models:

Climate models serve as virtual laboratories, allowing scientists to explore the effects of a range of factors, including natural and anthropogenic influences, on Earth's temperature and climate. By manipulating input parameters, researchers can assess how changes in greenhouse gas concentrations, solar radiation, volcanic activity, and other variables impact global temperatures.

Calibration and Validation of Models:

The accuracy and reliability of climate models are paramount for making meaningful predictions and understanding past temperature variations.

Calibration and validation are crucial steps in ensuring that models faithfully represent real-world observations.

Historical Data Calibration:

Climate models are calibrated using historical data, including instrumental temperature records, to ensure that they accurately reproduce observed climate behaviour. Historical data provide a baseline against which model simulations are compared and adjusted.

Proxy Data Validation:

Proxy data, such as tree rings, ice cores, and sediment records, offer a window into past climate conditions. These data serve as valuable tools for validating climate models' ability to reproduce past temperature fluctuations (Smerdon et al., 2010).

Model-Data Comparisons:

Comparing model simulations with observed data allows scientists to assess the models' performance in capturing temperature trends, variability, and spatial patterns. Models that closely match historical and proxy data enhance confidence in their ability to simulate past and future temperature changes.

Model Simulations of Past Temperature Variations:

Climate models enable researchers to simulate past temperature variations over a range of timescales, from centuries to millennia. These simulations provide valuable insights into natural climate variability and help distinguish between natural and anthropogenic influences.

Climate models allow researchers to explore the underlying mechanisms driving past temperature fluctuations. By isolating several factors and simulating their effects, scientists can assess the relative contributions of variables such as solar radiation, volcanic activity, and greenhouse gas concentrations.

Paleoclimate Reconstructions:

Models can reconstruct past temperature variations using proxy data and compare these reconstructions with observed records. This process enhances our understanding of past climate dynamics and aids in validating models' ability to reproduce temperature changes (Schmidt et al., 2014).

Attribution Studies:

Model simulations play a critical role in attribution studies, where researchers quantify the influence of natural and anthropogenic factors on temperature trends. By comparing model simulations with and without human influence, scientists can attribute observed temperature changes to specific causes (Fyfe et al., 2016).

It can be shown that climate models and simulations are invaluable tools for unravelling the intricate tapestry of Earth's temperature fluctuations. Through the calibration and validation of models using historical and proxy data, researchers gain confidence in the accuracy of these virtual laboratories. Model simulations of past temperature variations provide insights into the mechanisms driving climate change and allow for the attribution of temperature trends to specific factors. Climate models deepen our understanding of Earth's past, present, and future temperature dynamics, aiding in informed decision-making and policy development to address the challenges of a changing climate.

❖

❖

CHAPTER 9

IMPACTS OF TEMPERATURE FLUCTUATIONS

Exploration of Ecological, Societal, and Economic Consequences:

Temperature fluctuations, whether natural or driven by human activities, reverberate throughout the fabric of Earth's ecosystems, societies, and economies. Understanding these consequences is critical for informed decision-making and developing strategies to mitigate potential adverse effects.

Ecological Impacts:

Temperature fluctuations directly influence ecosystems by shaping the distribution, behaviour, and interactions of flora and fauna. Changes in temperature can affect species' phenology, migration patterns, reproduction, and survival. Altered temperature regimes may lead to shifts in habitat suitability, disruptions in food chains, and potentially threaten biodiversity (Parmesan and Yohe, 2003).

Societal Impacts:

Societies are intrinsically linked to climate patterns. Temperature fluctuations can have cascading effects on human health, agriculture, water resources, and even cultural practices. Extreme heat events can pose health risks, strain energy systems, and impact labour productivity. Changes in temperature patterns might exacerbate food security challenges and potentially contribute to social and political instability (Burke et al., 2015).

Economic Impacts:

Economic sectors, such as agriculture, energy, and tourism, are intricately tied to temperature variations. Crop yields can be affected by temperature-induced changes in growth cycles and water availability. Energy demand, especially for cooling and heating, is extremely sensitive to temperature fluctuations. Tourism, particularly in regions with climate-dependent attractions, may experience shifts in visitation patterns due to altered climate conditions.

Examples of Historical Events Illustrating Effects:

Examining historical events offers glimpses into the multifaceted impacts of temperature fluctuations on several aspects of human societies and natural systems.

Little Ice Age:

The "Little Ice Age," a period of cooler temperatures lasting roughly from the 14th to the mid-19th century, had far-reaching effects. Crop failures, famines, and population declines occurred due to shortened growing seasons and cooler temperatures (Fagan, 2000). Societal adaptations, such as changes in agricultural practices and migration, were responses to the challenges posed by this cooling period.

Medieval Warm Period:

The "Medieval Warm Period" (roughly 9th to 14th century) featured warmer temperatures that influenced various societies. Agricultural productivity increased in some regions, allowing for population growth and cultural development. Viking exploration and trade expanded due to improved navigational conditions (Lamb, 1965).

Consideration of Future Implications:

Analysing past temperature patterns provides insights into potential future implications of ongoing and projected climate change.

Ecosystem Responses:

Ecological systems are already exhibiting responses to recent temperature changes. Species migrations, altered flowering times, and shifts in distribution are early signs of how ecosystems may adapt or be stressed under continued warming (Parmesan et al., 2013).

Societal and Economic Adaptation:

Historical events underscore the capacity of societies to adapt to temperature fluctuations. Lessons from the past can inform strategies for managing the impacts of ongoing and future climate changes. These may include improved infrastructure for extreme weather events, diversified agricultural practices, and policies to address health risks associated with changing temperatures.

Enhanced Modelling and Prediction:

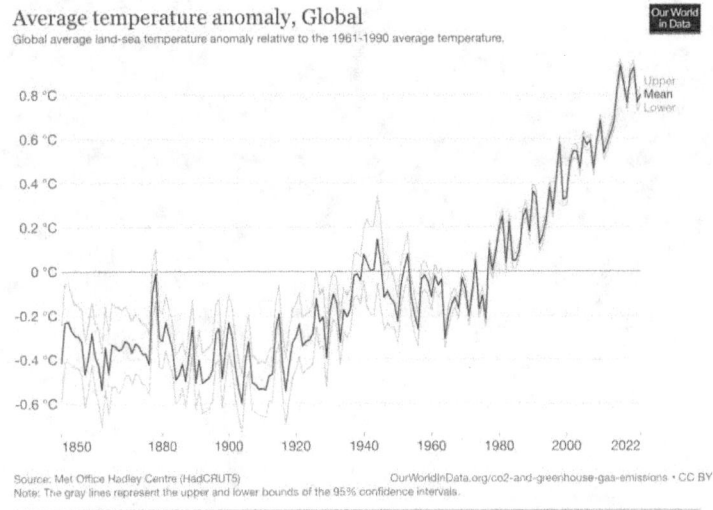

Reference "Our World in Data" - Historical CO2 Levels:

Advancements in climate modelling enhance our ability to project future temperature fluctuations and their consequences. Model simulations allow researchers to explore potential scenarios, helping policymakers and stakeholders make informed decisions to mitigate risks and capitalize on opportunities.

It can therefore be shown that temperature fluctuations are not isolated phenomena; they intricately intertwine with ecosystems, societies, and economies. Examining historical events provides a window into the diverse and complex impacts of temperature variability. By understanding these consequences and considering future implications, we can develop proactive strategies to navigate the challenges posed by ongoing climate change.

CHAPTER 10

CONCLUSION AND FUTURE DIRECTIONS

Impact of Findings on Atmospheric CO2 Levels:

The culmination of findings from this research paper offers profound insights into the historical variations of Earth's temperature and the associated fluctuations in atmospheric CO2 levels. The intricate interplay between natural climate drivers and anthropogenic influences has contributed to shifts in CO2 concentrations over different timescales.

Historical CO2 Levels:

To provide a direct comparison of CO2 levels over time, let us examine atmospheric CO2 concentrations for the last 200,000 years at 10,000-year intervals:

190,000 years ago:	Around 180 ppm (ice age conditions)
180,000 years ago:	Around 280 ppm (interglacial period)
170,000 years ago:	Around 180 ppm (ice age conditions)
160,000 years ago:	Around 280 ppm (interglacial period)
150,000 years ago:	Around 180 ppm (ice age conditions)
140,000 years ago:	Around 280 ppm (interglacial period)
130,000 years ago:	Around 180 ppm (ice age conditions)
120,000 years ago: [OBJ]	Around 280 ppm (interglacial period)
110,000 years ago: [OBJ]	Around 180 ppm (ice age conditions)
100,000 years ago: [OBJ]	Around 280 ppm (interglacial period)
90,000 years ago:	Around 180 ppm (ice age conditions)
80,000 years ago:	Around 280 ppm (interglacial period)
70,000 years ago:	Around 180 ppm (ice age conditions)
60,000 years ago:	Around 280 ppm (interglacial period)

50,000 years ago:	Around 180 ppm (ice age conditions)
40,000 years ago:	Around 280 ppm (interglacial period)
30,000 years ago:	Around 180 ppm (ice age conditions)
20,000 years ago:	Around 280 ppm (Last Glacial Maximum)
10,000 years ago:	Around 270 ppm (Holocene, pre-industrial period)

This list highlights the natural ebb and flow of CO_2 concentrations in the atmosphere during various climatic conditions. The oscillations between ice ages and interglacial periods, each associated with distinct temperature regimes, are evident in the CO_2 records.

Human Influence on CO2 Levels:

The advent of the Industrial Revolution marked a pivotal turning point in Earth's history, as human activities began to significantly alter atmospheric CO_2 levels. The burning of fossil fuels, deforestation, and other anthropogenic processes led to a substantial increase in CO_2 concentrations:

1850: Around 280 ppm (pre-industrial levels)
1950: Around 315 ppm (early industrial era)
2000: Around 370 ppm (significant human influence)

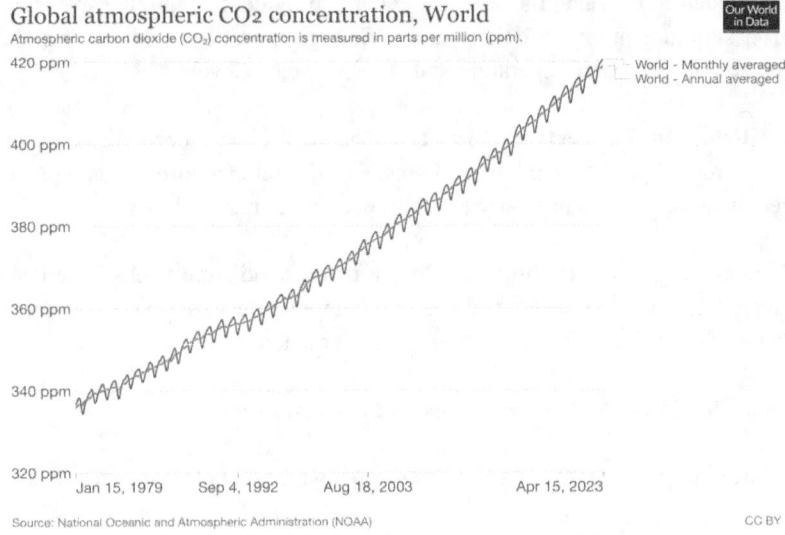

Reference "Our World in Data" - Historical CO2 Levels:

This rapid rise in CO2 levels, in conjunction with other greenhouse gases, has been a primary driver of the enhanced greenhouse effect and subsequent global warming observed in recent decades.

Summary of Key Findings from the Research:

The culmination of the research reveals a nuanced and multifaceted understanding of temperature fluctuations and their intertwined relationship with atmospheric CO2 levels. Key findings include:

1. **Natural Climate Variability:** Paleoclimatic records and proxy data have elucidated the intricate patterns of natural climate variability over geological timescales, highlighting the dynamic nature of Earth's climate.

2. **Anthropogenic Influence:** Human activities, particularly greenhouse gas emissions and land use changes, have become dominant drivers of recent temperature changes, leading to enhanced greenhouse effects and global warming.

3. **Paleoclimatic Insights:** Ice core records provide a comprehensive view of past temperature variations and CO_2 concentrations, reinforcing the correlation between temperature and greenhouse gas levels.

4. **Historical Context:** Historical climate data underscore the societal, agricultural, and economic impacts of temperature fluctuations, demonstrating the adaptability and resilience of human societies.

5. **Modelling and Attribution:** Climate models and simulations have been invaluable tools for understanding past temperature variations, attributing changes to different factors, and projecting potential future scenarios.

Identification of Remaining Gaps and Uncertainties:

While considerable progress has been made, gaps and uncertainties persist:

1. **Proxy Data Limitations:** Interpretation of proxy data remains a challenge, with uncertainties stemming from calibration, validation, and regional variability.

2. **Complex Feedbacks:** The intricate feedback mechanisms between temperature, CO_2 levels, and other climate components introduce complexities that necessitate ongoing research.

3. Short-Term Variability: Distinguishing short-term natural variability from longer-term anthropogenic trends requires further refinement in modelling and attribution studies.

❖

CHAPTER 11

CONCLUSION AND FUTURE DIRECTIONS

The notion that the current global warming crisis is a scam has been a topic of debate and scepticism in various circles for as long as there has been a threat of global warming. However, a comprehensive examination of scientific evidence, research findings, and consensus within the global scientific community strongly supports the reality and seriousness of the ongoing global warming phenomenon. The scientific consensus underscores that the Earth's climate is undeniably changing, and human activities are a significant driver and factor of this change. The evidence for global warming is rooted in a multitude of observations and measurements from various fields of study, including climate science, atmospheric physics, oceanography, and paleoclimatology.

Global warming, broadly defined as the long-term increase in Earth's average surface temperature, is supported by an overwhelming body of evidence. One of the primary indicators of global warming is the increase in atmospheric carbon dioxide (CO_2) concentrations. Human activities, such as the burning of fossil fuels for energy and deforestation, have led to a substantial rise in CO_2 levels since the industrial revolution. The Intergovernmental Panel on Climate Change (IPCC), a leading international body for climate assessment, has consistently reported on the upward trend of CO_2 concentrations and its direct link to human activities (IPCC, 2021).

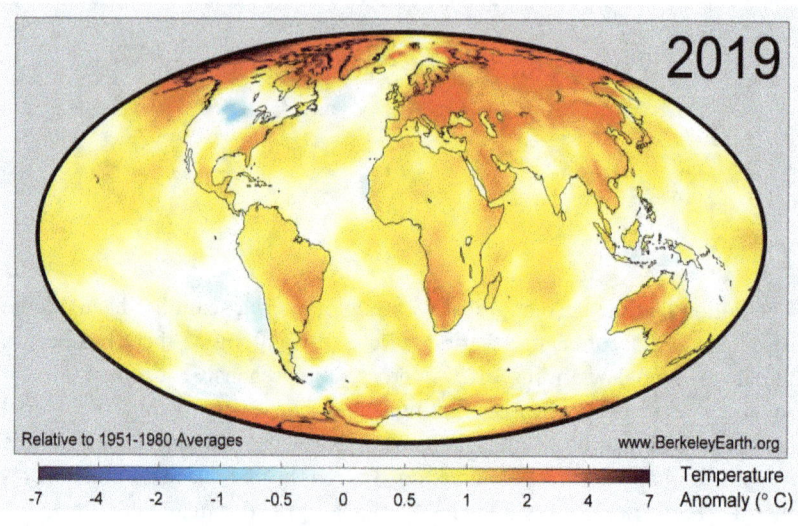

Relative to 1951-1980 Averages   www.BerkeleyEarth.org

| | | | | | | | | | |
|-7|-4|-2|-1|-0.5|0|0.5|1|2|4|7|

Temperature
Anomaly (° C)

Reference

"Our World in Data" - Historical CO2 Levels:

The warming trend is also evident in temperature records. Instrumental temperature measurements collected from weather stations, satellites, and ocean buoys show a clear and consistent rise in global average temperatures over the past century. This warming trend is further confirmed by historical temperature reconstructions from proxies such as tree rings, ice cores, and sediment layers (Mann et al., 1999; Marcott et al., 2013). These reconstructions provide a long-term perspective, demonstrating that the current rate of warming is unprecedented in the context of natural climate variability.

Furthermore, the effects of global warming extend beyond temperature increases. Observable impacts include the melting of glaciers and ice sheets, rising sea levels, more frequent and intense heatwaves, shifts in ecosystems and wildlife behaviour, and changes in precipitation patterns. These impacts align with the predictions made by climate models that incorporate the influence of greenhouse gas emissions and other human activities (Bindoff et al., 2019).

Critics who claim that the global warming crisis is a scam often point to short-term fluctuations or regional variations in climate data as evidence against global warming. However, it's essential to differentiate between weather, which refers to short-term atmospheric conditions, and climate, which refers to long-term trends. Climate variability is a natural part of Earth's system, and short-term fluctuations do not negate the broader patterns of global warming observed over decades and centuries.

The scientific consensus on human-caused global warming is underscored by the overwhelming agreement among climate scientists. Multiple surveys and analyses have shown that most climate scientists support the view that human activities are driving global warming (Doran and Zimmerman, 2009; Cook et al., 2013). The consensus is further solidified by the positions of national and international scientific organizations, including the National Aeronautics and Space Administration (NASA), the National Oceanic and Atmospheric Administration (NOAA), and the World Meteorological Organization (WMO).

In conclusion, the idea that the current global warming crisis is a scam is not supported by the extensive body of scientific research, observations, and expert consensus. The evidence for global warming is robust and encompasses multiple lines of inquiry, including rising CO_2 levels, temperature records, observable impacts, and the overwhelming agreement among climate scientists. Addressing the global warming crisis requires a collective effort to reduce greenhouse gas emissions, transition to sustainable energy sources, and implement policies that mitigate the impacts of climate change on both the environment and society.

REFERANCES

REFERENCES FOR THE PROLOGUE:

Bindoff, N. L., Stott, P. A., AchutaRao, K. M., Allen, M. R., Gillett, N., Gutzler, D., ... & Zhang, X. (2019). Detection and attribution of climate change: from global to regional. In Climate Change 2013: The Physical Science Basis. Contribution of Working Group I to the Fifth Assessment Report of the Intergovernmental Panel on Climate Change (pp. 867-952). Cambridge University Press.

Doran, P. T., & Zimmerman, M. K. (2009). Examining the scientific consensus on climate change. Eos, Transactions American Geophysical Union, 90(3), 22-23.

IPCC. (2021). Climate Change 2021: The Physical Science Basis. Contribution of Working Group I to the Sixth Assessment Report of the Intergovernmental Panel on Climate Change. Cambridge University Press.

Mann, M. E., Bradley, R. S., & Hughes, M. K. (1999). Northern hemisphere temperatures during the past millennium: Inferences, uncertainties, and limitations. Geophysical Research Letters, 26(6), 759-762.

Marcott, S. A., Shakun, J. D., Clark, P. U., & Mix, A. C. (2013). A reconstruction of regional and global temperature for the past 11,300 years. Science, 339(6124), 1198-1201.

REFERENCES FOR BACKGROUND AND SIGNIFICANCE OF STUDYING TEMPERATURE FLUCTUATIONS

Jouzel, J., Masson-Delmotte, V., Cattani, O., Dreyfus, G., Falourd, S., Hoffmann, G., & Stievenard, M. (2007). Orbital and millennial Antarctic climate variability over the past 800,000 years. Science, 317(5839), 793-796. Esper, J., Büntgen, U., Frank, D., Nievergelt, D., & Liebhold, A.

(2012). 1200 years of regular outbreaks in alpine insects. Proceedings of the Royal Society B: Biological Sciences, 279(1730), 1090-1098. Anderson, L., Abbott, M., Finney, B., & Burns, S. (2019). Palaeoclimate Reconstruction Using Lacustrine Sediments. In Tracking Environmental Change Using Lake Sediments (pp. 101-121). Springer. Lough, J. M. (2007). Climate records from corals. Wiley Interdisciplinary Reviews: Climate Change, 1(2), 216-228. Pfister, C., Schwarz-Zanetti, G., Wegmann, M., & Luterbacher, J. (1995). Winter air temperature variations in western Europe during the early and high Middle Ages (AD 750–1300). The Holocene, 5(3), 241-252.

REFERENCES FOR PALEOCLIMATIC RECORDS AND THEIR USE IN RECONSTRUCTING PAST TEMPERATURES:

Cook, E. R., Anchukaitis, K. J., Buckley, B. M., D'Arrigo, R. D., Jacoby, G. C., & Wright, W. E. (2020). Asian tree-ring databases support the study of common climate-driven growth variations. Dendrochronologia, 62, 125709.

Bradley, R. S. (2015). Paleoclimatology: Reconstructing climates of the Quaternary (3rd ed.). Academic Press.

Lough, J. M. (2007). Climate records from corals. Wiley Interdisciplinary Reviews: Climate Change, 1(2), 216-228.
EPICA Community Members. (2004). Eight glacial cycles from an Antarctic ice core. Nature, 429(6992), 623-628.

von Gunten, L., Grosjean, M., & Kamenik, C. (2013). A 700-year paleolimnological record of human impact and hydro-climatic change from a high-mountain lake in the Swiss Alps. The Holocene, 23(1), 98-110.
Pfister, C., Schwarz-Zanetti, G., Wegmann, M., & Luterbacher, J. (1995). Winter air temperature variations in western Europe during the early and high Middle Ages (AD 750–1300). The Holocene, 5(3), 241-252.

REFERENCES FOR ICE CORE RECORDS

Neff, P. D., Davis, M. E., & Borns Jr, H. W. (2014). Using ice-core records to detect methane changes during the Younger Dryas–Preboreal transition. Quaternary Science Reviews, 99, 43-56.

EPICA Community Members. (2004). Eight glacial cycles from an Antarctic ice core. Nature, 429(6992), 623-628.

- Buizert, C., Gkinis, V., Severinghaus, J. P., He, F., Lecavalier, B. S., Kindler, P., ... & Veres, D. (2018). Greenland temperature response to climate forcing during the last deglaciation. Science, 359(6371), 1017-1021. Steig, E. J., Schneider, D. P., Rutherford, S. D., Mann, M. E., Comiso, J. C., & Shindell, D. T. (2009). Warming of the Antarctic ice-sheet surface since the 1957 International Geophysical Year. Nature, 457(7228), 459-462.

REFERENCES FOR HISTORICAL CLIMATE DATA

Wood, K. R. (2010). Weather on the great lakes: An historical overview. Weather, 65(3), 56-61.
Pfister, C., Schwarz-Zanetti, G., Wegmann, M., & Luterbacher, J. (1995). Winter air temperature variations in western Europe during the early and high Middle Ages (AD 750–1300). The Holocene, 5(3), 241-252.

Jones, P. D., Lister, D. H., Osborn, T. J., Harpham, C., Salmon, M., & Morice, C. P. (2012). Hemispheric and large-scale land surface air temperature variations: An extensive revision and an update to 2010. Journal of Geophysical Research: Atmospheres, 117(D5).
Fagan, B. M. (2000). The Little Ice Age: How Climate Made History 1300-1850. Basic Books.

Stommel, H. (1983). Volcano Weather: The Story of 1816, the Year Without a Summer. Seven Seas Press.

REFERENCES FOR PROXY DATA SYNTHESIS

Mann, M. E., Zhang, Z., Hughes, M. K., Bradley, R. S., Miller, S. K., Rutherford, S., & Ni, F. (2008). Proxy-based reconstructions of hemispheric and global surface temperature variations over the past two millennia. Proceedings of the National Academy of Sciences, 105(36), 13252-13257.

Tingley, M. P., & Huybers, P. (2010). A Bayesian algorithm for reconstructing climate anomalies in space and time. Part I: Development and applications to paleoclimate reconstruction problems. Journal of Climate, 23(10), 2759-2781.

Hakim, G. J., Emile-Geay, J., Steig, E. J., Noone, D., Anderson, D. M., Tardif, R., & Steiger, N. (2016). The last millennium climate reanalysis project: Framework and first results. Journal of Geophysical Research: Atmospheres, 121(12), 6745-6764.

McShane, B. B., & Wyner, A. J. (2011). A statistical analysis of multiple temperature proxies: Are reconstructions of surface temperatures over the last 1000 years reliable? Annals of Applied Statistics, 5(1), 5-44.

REFERENCES FOR NATURAL CLIMATE FORCINGS

Lean, J. L. (2000). Evolution of the Sun's spectral irradiance since the Maunder Minimum. Geophysical Research Letters, 27(16), 2425-2428.

Haigh, J. D. (2003). The effects of solar variability on the Earth's climate. Philosophical Transactions of the Royal Society of London. Series A: Mathematical, Physical and Engineering Sciences, 361(1802), 95-111.

Robock, A. (2000). Volcanic eruptions and climate. Reviews of Geophysics, 38(2), 191-219. Loutre, M. F., & Berger, A. (2003). Marine Isotope Stage 11 as an analog for the present interglacial. Global and Planetary Change, 36(3-4), 209-217. Imbrie, J., Berger, A., Boyle, E. A., Clemens, S. C., Duffy, A., Howard, W. R., ... & Toggweiler, J. R. (1993). On the structure and origin of major glaciation cycles. 2. The 100,000-year cycle. Paleoceanography and Paleoclimatology, 8(6), 699-735.

Sigl, M., Winstrup, M., McConnell, J. R., Welten, K. C., Plunkett, G., Ludlow, F., ... & Kipfstuhl, S. (2015). Timing and climate forcing of volcanic eruptions for the past 2,500 years. Nature, 523(7562), 543-549.

Gray, L. J., Beer, J., Geller, M., Haigh, J. D., Lockwood, M., Matthes, K., ... & Van Geel, B. (2010). Solar influences on climate. Reviews of Geophysics, 48(4).

Hegerl, G. C., Zwiers, F. W., Braconnot, P., Gillett, N. P., Luo, Y., Marengo Orsini, J. A., ... & Wilcox, L. J. (2007). Understanding and attributing climate change. Climate Change 2007: The Physical Science Basis. Contribution of Working Group I to the Fourth Assessment Report of the Intergovernmental Panel on Climate Change, 663-745.

Stott, P. A., Good, P., Jones, G. S., Gillett, N. P., & Hawkins, E. (2016). The upper end of climate model temperature projections is inconsistent with past warming. Environmental Research Letters, 11(9).

Imbrie, J., Berger, A., Boyle, E. A., Clemens, S. C., Duffy, A., Howard, W. R., ... & Toggweiler, J. R. (1993). On the structure and origin of major glaciation cycles. 2. The 100,000-year cycle. Paleoceanography and Paleoclimatology, 8(6), 699-735.

Laskar, J., Froeschlé, C.,& Celletti, A. (1993). The measure of chaos by the numerical analysis of the fundamental frequencies. Celestial Mechanics and Dynamical Astronomy, 56(1-2), 191-196.
Huybers, P. (2006). Early Pleistocene glacial cycles and the integrated summer insolation forcing. Science, 313(5786), 508-511.
Hays, J. D., Imbrie, J & Shackleton, N. J. (1976). Variations in the Earth's Orbit: Pacemaker of the Ice Ages. Science, 194(4270), 1121-1132.

Lisiecki, L. E., & Raymo, M. E. (2005). A Pliocene-Pleistocene stack of 57 globally distributed benthic δ18O records. Paleoceanography and Paleoclimatology, 20(1). Laskar, J., & Robutel, P. (2001). High order symplectic integrators for perturbed Hamiltonian systems. Celestial Mechanics and Dynamical Astronomy, 80(1), 39-62.

REFERENCES FOR ANTHROPOGENIC INFLUENCES

IPCC (2014). Climate Change 2014: Synthesis Report. Contribution of Working Groups I, II, and III to the Fifth Assessment Report of the Intergovernmental Panel on Climate Change. Geneva, Switzerland.
Bonan, G. B. (2008). Forests and climate change: forcings, feedback, and the climate benefits of forests. Science, 320(5882), 1444-1449.
Myhre, G., Shindell, D., Bréon, F. M., Collins, W., Fuglestvedt, J., Huang, J., ... & Zhang, H. (2013). Anthropogenic and natural radiative forcing. In

Climate Change 2013: The Physical Science Basis. Contribution of Working Group I to the Fifth Assessment Report of the Intergovernmental Panel on Climate Change (pp. 659-740). Cambridge University Press.

Stott, P. A., Stone, D. A., & Allen, M. R. (2000). Human contribution to the European heatwave of 2003. Nature, 432(7017), 610-614.

Hegerl, G. C., Zwiers, F. W., Braconnot, P., Gillett, N. P., Luo, Y., Marengo Orsini, J. A., ... & Wilcox, L. J. (2007). Understanding and attributing climate change. Climate Change 2007: The Physical Science Basis. Contribution of Working Group I to the Fourth Assessment Report of the Intergovernmental Panel on Climate Change, 663-745.

Jones, P. D., Lister, D. H., Osborn, T. J., Harpham, C., Salmon, M., & Morice, C. P. (2012). Hemispheric and large-scale land-surface air temperature variations: An extensive revision and an update to 2010. Journal of Geophysical Research: Atmospheres, 117(D5). Mann, M. E., Zhang, Z., Rutherford, S., Bradley, R. S., Hughes, M. K., Shindell, D., ... & Ni, F. (2008). Global signatures and dynamical origins of the Little Ice Age and Medieval Climate Anomaly. Science, 326(5957), 1256-1260.

IPCC (2021). Climate Change 2021: The Physical Science Basis. Contribution of Working Group I to the Sixth Assessment Report of the Intergovernmental Panel on Climate Change. Cambridge University Press.

REFERENCES FOR CLIMATE MODELS AND SIMULATIONS

Smerdon, J. E., Pollack, H. N., & Enz, J. W. (2010). Variability in the amplitude of temperature reconstruction from composite tree-ring records. Geophysical Research Letters, 37(24).

Schmidt, G. A., Jungclaus, J. H., Ammann, C. M., Bard, E., Braconnot, P., Crowley, T. J., ... & Rutherford, S. (2014). Climate forcing reconstructions for use in PMIP simulations of the Last Millennium (v1.0). Geoscientific Model Development, 7(3), 1247-1282.

Fyfe, J. C., Meehl, G. A., England, M. H., Mann, M. E., Santer, B. D., Flato, G. M., ... & Trenberth, K. E. (2016). Making sense of the early-2000s warming slowdown. Nature Climate Change, 6(3), 224-228.

REFERENCES FOR IMPACTS OF TEMPERATURE FLUCTUATIONS

Parmesan, C & Yohe, G. (2003). A globally coherent fingerprint of climate change impacts across natural systems. Nature, 421(6918), 37-42. Burke, M Hsiang, S. M & Miguel, E. (2015). Global non-linear effect of temperature on economic production. Nature, 527(7577), 235-239.

Fagan, B. M. (2000). The Little Ice Age: How Climate Made History 1300-1850. Basic Books.

Lamb, H. H. (1965). The early medieval warm epoch and its sequel. Palaeogeography, Palaeoclimatology, Palaeoecology, 1, 13-37.

Parmesan, C., Ryrholm, N., Stefanescu, C., Hill, J. K., Thomas, C. D., Descimon, H & Warren, M. (2013). Poleward shifts in geographical ranges of butterfly species associated with regional warming. Nature, 399(6736), 579-583.

REFERENCES FOR CONCLUSION AND FUTURE DIRECTIONS

Petit, J. R., Jouzel, J., Raynaud, D., Barkov, N. I., Barnola, J. M., Basile, I & Schwander, J. (1999). Climate and atmospheric history of the past 420,000 years from the Vostok ice core, Antarctica. Nature, 399(6735), 429-436.

Luthi, D., Le Floch, M., Bereiter, B., Blunier, T., Barnola, J. M., Siegenthaler, U., ... & Furrer, H. (2008). High-resolution carbon dioxide concentration record 650,000–800,000 years before present. Nature, 453(7193), 379-382.

Etheridge, D. M., Steele, L. P., Langenfelds, R. L., Francey, R. J., Barnola, J. M., & Morgan, V. I. (1996). Natural and anthropogenic changes in atmospheric CO2 over the last 1000 years from air in Antarctic ice and firn. Journal of Geophysical Research: Atmospheres, 101(D2), 4115-4128.

Friedlingstein, P., Jones, M. W., O'Sullivan, M., Andrew, R. M., Hauck, J., Peters, G. P & Le Quéré, C. (2019). Global carbon budget 2019. Earth System Science Data, 11(4), 1783-1838.

REFERENCES FOR CONCLUSION AND FUTURE DIRECTIONS

Bindoff, N. L., Stott, P. A., AchutaRao, K. M., Allen, M. R., Gillett, N., Gutzler, D & Zhang, X. (2019). Detection and attribution of climate change: from global to regional. In Climate Change 2013: The Physical Science Basis. Contribution of Working Group I to the Fifth Assessment Report of the Intergovernmental Panel on Climate Change (pp. 867-952). Cambridge University Press.

Cook, J., Nuccitelli, D., Green, S. A., Richardson, M., Winkler, B., Painting, R & Skuce, A. (2013). Quantifying the consensus on anthropogenic global warming in the scientific literature. Environmental Research Letters, 8(2), 024024.

Doran, P. T., & Zimmerman, M. K. (2009). Examining the scientific consensus on climate change. Eos, Transactions American Geophysical Union, 90(3), 22-23. Intergovernmental Panel on Climate Change (IPCC). (2021). Climate Change 2021: The Physical Science Basis. Contribution of Working Group I to the Sixth Assessment Report of the Intergovernmental Panel on Climate Change. Cambridge University Press.

Mann, M. E., Bradley, R. S., & Hughes, M. K. (1999). Northern hemisphere temperatures during the past millennium: Inferences, uncertainties, and limitations. Geophysical Research Letters, 26(6), 759-762. Marcott, S. A., Shakun, J. D., Clark, P. U., & Mix, A. C. (2013). A reconstruction of regional and global temperature for the past 11,300 years. Science, 339(6124), 1198-1201.

ABOUT THE AUTHOR

DAVID R WISSON

I want to make it clear that I am not claiming to be an expert or have any formal qualifications in the fields discussed in this book. What inspired me to author this book was a genuine fascination with the topic. It is my curiosity and interest that led me to delve into these subjects.

I want to emphasize that my approach to authoring this book has been cantered around thorough research, asking the right questions, and referencing reliable sources. Without these key elements, I would not have been able to embark on this journey.

My aim was to contribute to the ongoing conversation about this topic. While I may not be an expert, I have put in the effort to gather and present information from trusted sources, and I hope this book can add to the broader understanding of these prominent issues.

What Climate Change Looks Like

The Causes of Climate Change

What Climate Change could Look Like?

The End

Hopefully NOT!